The best of
Warbirds Over Wanaka

Aerial scenes of a show. From these images taken from 1988 through 2000, one can see how dramatically both the airport and crowd numbers have grown.

The best of Warbirds Over Wanaka

Text by Ian Brodie
Air-to-air photography by Ian Brodie and Phil Makanna

This book is dedicated to all the staff and volunteers who, every two years, forgo all normality of life to present Warbirds Over Wanaka.

Established in 1907, Reed Publishing (NZ) Ltd is New Zealand's largest book publisher, with over 300 titles in print.

For details on all these books visit our website: **www.reed.co.nz**

Published by Reed Books, a division of Reed Publishing (NZ) Ltd, 39 Rawene Rd, Birkenhead, Auckland 10. Associated companies, branches and representatives throughout the world.

This book is copyright. Except for the purpose of fair reviewing, no part of this publication may be reproduced or transmitted in any form or by any means, electronic or mechanical, including photocopying, recording, or any information storage and retrieval system, without permission in writing from the publisher. Infringers of copyright render themselves liable to prosecution.

© 2002 Alpine Deer Group
The authors assert their moral rights in the work.

ISBN 07900 837 8
First published 2002

Designed by Helen Barker
Edited by Carolyn Lagahetau

Printed in New Zealand

Contents

Introduction 7
The shows
1988 18
1990 23
1992 29
1994 39
1996 50
1998 66
2000 78

Geoff Sloan

Introduction *by Ian Brodie*

The essence of Warbirds Over Wanaka: vintage, veteran and classic aircraft combine with military vehicles and symbols of our agricultural past to present entertainment for the whole family.

Warbirds Over Wanaka is a name that evokes instant worldwide recognition as one of the best warbird airshows to see and be involved with. In just fourteen years this airshow has grown in stature from a small country fair to an event that draws over 100,000 people every second Easter to the small airport situated on the outskirts of Wanaka.

Central Otago seems an unlikely place to hold an event of this nature, but if one explores the airshow's beginnings and development, a better understanding can be made as to why Wanaka is the perfect place for Warbirds Over Wanaka to take place.

A good idea?

It is 1987 and Tim Wallis sits in his Alpine Deer Group office at Wanaka Airport. He has an idea — to hold an airshow at the local airport and maybe attract a few thousand people. Tim has an unbridled passion for aviation that he loves to share. His dream of gathering a collection of the greatest fighters of World War II is gathering momentum. Having sold his P-51D Mustang to allow the purchase of a Supermarine Spitfire, he is also looking at various other warbirds, including a Curtiss Kittyhawk and a Vought Corsair — aircraft that have not been seen nor heard in New Zealand skies for over 40 years.

Rather than just aeroplanes, Tim decides to make the day a celebration of bygone eras in New Zealand by incorporating vintage tractors and machinery into the show. It will essentially be a day of living history, allowing people to not only see but hear, smell and touch these links with the past. With his brother George, an avid collector of agricultural equipment, the two talk many of their friends into gathering exhibits for ground and air displays.

Tim calls the local Lions Club, shares his idea, and asks for their assistance in organising a flying display to be held in January 1988, which will be called Warbirds On Parade. After some discussion within the local community it is agreed that Easter would be a better time for the display: Wanaka is filled with holidaymakers in January regardless of what is going on and the club feels the show would better benefit the community if held during a quieter time.

Tim talks to members of the New Zealand Warbirds Association in Auckland and they agree to fly down for the weekend with a number of aircraft.

1988 — the first show

In 1988 Wanaka Airport looked considerably different than it does today. The runway was sealed only the year previously and the Alpine Deer Group Ltd head office, with its small attached hangar, and Aspiring Air were the only permanent residents. There was plenty of space to park the vintage machinery and vehicles and a 1200 metre runway allowed a long display line for the crowd.

I was working for Air New Zealand at the time and convinced management that we should take a Fokker Friendship filled with enthusiasts to the show for the day. The flight sold out in a week, showing there was definite public interest in the display. Tim's enthusiasm and a lot of hard work by many others ensured the weekend was a success. Much to the delight of the organising team, Warbirds on Parade attracted 14,000 visitors. It was not just the aircraft that the visitors enjoyed — they were also entertained with displays of vintage farm machinery, tractors and stationary engines. It was this unique country fair atmosphere, combined with flying warbirds, that captured the public's enthusiasm. Highlights included visits and displays from a de Havilland Venom, a Hawker Sea Fury and the wonderful Harvards that made up the Roaring Forties team.

The North American T-6 Harvard is powered by a 600 hp radial engine, and has provided the backbone for New Zealand's warbird movement. The aircraft's unique sound is made with the propeller-tips breaking the sound barrier. Here, nine are in formation during the 1992 show.
David Evans

There are always aircraft that make it to the show only through the sheer hard work of the owners and engineers undertaking a rebuild, often over a number of years. This picture shows Ray Mulqueen, Chief Engineer for the Alpine Fighter Collection, overseeing the first engine run since rebuild of a Curtiss P-40K Kittyhawk — this on the Thursday before Easter 1992! *Ian Brodie*

Inviting pilots who flew warbirds 'in anger' is another important aspect of Warbirds Over Wanaka and provides a direct link with the past. Here, Sir Tim (left) interviews Rusty Leith, who in 1945 flew the same Spitfire now owned by Sir Tim while with 453 Squadron, based in England. Watching are New Zealand fighter pilot Ray Archibald and Wayne Parsons (right), who has commentated at every airshow. *Alpine Fighter Collection*

In 1992 the appearance of the Chance Vought Corsair provided one half of a Royal New Zealand Air Force (RNZAF) Pacific fighter duo. *Ian Brodie*

In the small programme the following note appeared: *Should the combination of civil and warbird aircraft, vintage and veteran farm machinery and equipment be a success, it is hoped that it will become an annual or biennial event. Wanaka is a great venue for such an occasion.*

1990 — With a sneaky one in 1989!

Thrilled with the success of the event Tim decided on a biennial airshow. At the close of the 1988 airshow the next display, now renamed Warbirds Over Wanaka, was being planned for Easter 1990.

However, Tim found it hard to wait another two years. He organised a small fly-in to Wanaka over the Easter weekend of 1989. Again, members of the New Zealand

Warbirds Association attended, enjoying a relaxing weekend of flying and Tim's famous southern hospitality.

When the weekend concluded it was decided that a committee should be formed to look after the myriad details for the forthcoming 1990 show. It was decided that three members would comprise this committee: Tim, his brother George and Gavin Johnston, a local Lions Club member who had helped in 1988.

At the same time Tim's own collection of aircraft was growing. He had acquired a Curtiss P-40K Kittyhawk that was being rebuilt, and a shiny Supermarine Spitfire that would be displayed at Wanaka for the first time by pilot Stephen Grey.

The 1990 show attracted 28,000 people. News of the 1988 show had spread and visitors included a number of people from overseas. These included Paul Coggan, editor of *Warbirds Worldwide* and Pilot Officer Rusty Leith from Australia. Rusty, a wartime Spitfire pilot with 453 Squadron RAAF, was very keen to see Tim's Spitfire as it was the same Spitfire he had flown during World War II.

As well as the Spitfire, other highlights of the 1990 show included an aerobatic display in formation by a Venom and a Mustang and a BAC Strikemaster.

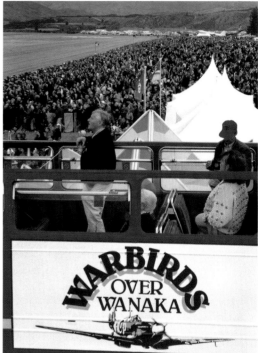

Sir Tim atop the familiar double-decker bus, 1994. *Shar Devine*

The Warbirds Over Wanaka committee: (left to right) Gavin Johnston, Sir Tim Wallis and George Wallis.
Ian Brodie

1992 — The year of the Me-109

To many New Zealanders, 1992 was 'The year of the Me-109'. The organising committee had realised that the show was going to grow and they needed to think of new and different displays. They decided to bring a Messerschmitt Me-109J from England for the Warbirds Over Wanaka 1992 show. Partially sponsored by the sale of 'Gold Pass' stand tickets, this aircraft and the promised dogfight with the Alpine Spitfire captured the public's imagination prior to the show and helped ensure an attendance of over 55,000 people. I remember taking a week's leave from Air New Zealand to be at Wanaka when the 109 arrived. The sight of it in formation with the Spitfire and the Alpine Fighter Collection's newly arrived Corsair in Wanaka's late evening light is one that will remain with me for ever. The subsequent display at the show saw Ray Hanna flying the Spitfire and his son, Mark, flying the Messerschmitt. This father and son team produced an amazing display.

Alpine Fighter Collection engineers had also been desperately finalising a project that was to see four years of work come to fruition. On Easter Saturday, just prior to the start of the airshow, the Kittyhawk, made her first post-restoration flight, and subsequently displayed on schedule that afternoon in the hands of Mark Hanna.

1994 — The 'sinking' of the *Yamato*

By 1994 the reputation, format and excitement of Warbirds Over Wanaka was well established. Tim and his team kept coming up with wonderful ideas and more and more warbirds had appeared on the New Zealand scene. What had been bar talk and a dream was now a reality. The New Zealand aviation trade was also attracting more and more interest. The Friday 'practice' day had become an Aviation Trade Day, with over 100 exhibitors displaying and selling their wares to the public. Warbirds Over Wanaka was now a three-day extravaganza, with a full flying display taking place on Saturday and Sunday.

On a personal note, 1994 was the year Warbirds Over Wanaka came of age. The professionalism of the pilots, organisers and participants covering all aspects of the weekend had helped to produce an event that justified an enormous sense of pride. I had joined Alpine Deer Group Ltd in mid-1992, so this was my first airshow 'from the inside'. I recall Tim asking me about a fortnight before Easter, 'What famous naval battles took place 50 years ago?' We came across the sinking of the *Yamato*. Within a week we had a 200-foot canvas profile of the ship completed for a mock sinking! I have to admit to being amazed and inspired by Tim's enthusiasm and passion.

The father and son team of Ray and Mark Hanna has been an integral part of Warbirds Over Wanaka. In September 1999 Mark was tragically killed in a flying accident in Spain. *Shar Devine*

Highlights of the weekend included the caravan drop — one way to rid the airport of an errant parker! We were privileged to receive a visit and consequent display of the famous Britten motorbike and announce the rebuild of a Hurricane Mk IIa, which was on display at the Fighter Pilots Museum.

Buoyed with the success of the airshow, Tim had been thinking for some time about holding smaller events at other times of the year. His thoughts were simple: a small fly-in would allow pilots to practise for the 'big event' and a minimal gate charge would cover the costs of fuel. The first of these was held in October 1994, in conjunction with a fighter pilots' reunion held at the New Zealand Fighter Pilots Museum. In January 1995 a further event was held to celebrate the first flight of the Yakovlev Yak-3M. Last-minute hitches were to delay the planned first display, but the small crowd enjoyed the sound of the engine at full power on a ground-run. The aircraft was displayed surrounded by snow (specially flown in) and some very appropriate bottles of Russian vodka!

On 19 March Tim continued his Russian theme in a display entitled AFC in Action and in May he took the warbirds to Christchurch. The importance of the fiftieth anniversary of the end of World War II in Europe did not escape him. He arranged to

Gavin Johnston 'at work' prior to an airshow.
Ian Brodie

Dick Thurman, the owner of this P-40, flew her from Auckland to attend the show in 2000.
Phil Makanna

celebrate and commemorate those who had been involved in the war with a special airshow and gala. All returned servicemen and servicewomen were given free admission to the Wigram show and they were entertained free of charge in a large marquee. This was typical of Tim's generosity and gratitude towards those that have served. The show was a huge success for those Tim intended to honour, but attendance by the general public was lower than expected.

1996 — Tim's absence

Tim was unable to attend the 1996 Warbirds Over Wanaka as he had been seriously injured in a take-off accident in his Mk XIV Spitfire during January. His absence during the all-important planning stages for the weekend was keenly felt by all, and it was strange not seeing him at the top of the double-decker bus that the commentary was run from.

However, the show attracted over 65,000 people and Tim did see some of the show from a nearby hill, enjoying every moment.

Displays included an art exhibition in the New Zealand Fighter Pilots Museum, the official welcome of the Yak-3 to the Alpine Fighter Collection's stable of aircraft, and a Daimler Benz-powered Messerschmitt Me-109. Mark Jefferies jollied everyone along when he painted a giant smiley face in the sky with his Yak-50.

1998 — Red Stars Rising

In the five years preceding 1998, deep in Siberia, a restoration process had been undertaken that led us to expect the arrival of six Polikarpov I-16 fighters at Wanaka. It was a natural and logical conclusion that Warbirds Over Wanaka 1998 was given the by-line of 'Red Stars Rising'. The first of these chubby little fighters arrived the weekend after the 1996 show. By early 1998 they were ready for their re-introduction to the aviation world; 65 years had passed since their debut.

Another special event was also to take place that Easter — the appearance of the Wanaka-based P-51D Mustang in a special colour scheme. Only one New Zealander was known to have flown in combat with the US 8th Army Air Force during World War II: Flight Lieutenant Jack Cleland. Jack had passed away but in 1998 his wife and family would be special guests at the show. It was appropriate to paint the Mustang in the colours of the aircraft he had flown in 1944. Interestingly, Cleland had flown several long-range escort missions that had necessitated landings in Russia to refuel — another Russian connection!

Another special guest at the show was one of Cleland's squadron comrades — Chuck Yeager.

News of the flight of the Polikarpovs had spread worldwide. Over 85,000 people attended the show and enjoyed seeing five Polikarpov I-16s in the air, an occurrence not seen since the late 1940s. This single event put Warbirds Over Wanaka at the forefront of international airshows.

'Warden Hodges' (Phil Murray) did his best to steal an aeroplane in 1998. However, the Dad's Army team saved the day. *Geoff Sloan*

The Russian theme also permeated other aspects of the show. There was Russian music accompanied by Russian dancers and the Warhorses at Wanaka team went all out to continue the illusion doing guard duty over the little planes. The guard-house (complete with Russian-speaking guards) provided many with the opportunity to experience a light-hearted Russian 'lock-up'.

2000 — The Hurricane is completed

Engineers at Hawker Restorations in the UK and Air New Zealand Engineering Services in Christchurch had worked for almost six years to restore the star of the 2000 show to her former glory. The aircraft was no stranger to Wanaka, having visited in 1994 as a collection of parts and in 1996 as a bare skeleton. The sixtieth anniversary year of the Battle of France and Battle of Britain was a most appropriate time for the debut of the Alpine Fighter Collection's Hawker Hurricane Mk IIA.

There was also no doubt that this show was to continue the Russian theme established in 1998. Three Polikarpov I-153 biplane fighters had arrived from Russia and were ready for their debut, accompanied, of course, by the I-16s that were introduced to the public in 1998.

In 2000 Warbirds Over Wanaka broke all attendance figures. It seemed that word had spread that the Polikarpov aircraft had to be seen to be believed. The phone of general manager Gavin Johnston rang continuously. Gold Passes sold out in record time and at the end of the show over 100,000 people had walked through the gates of Wanaka Airport.

A sea of gold. The Gold Pass area was introduced in 1992. Since then it has become the most popular place to be, with seats sold out well before Easter. *Alpine Fighter Collection*

Aerobatic displays don't come better than this. Nigel Arnot in his Sukhoi SU-31 in 1998. *Geoff Sloan*

They came and they saw — not only a Hurricane but eight Polikarpov fighters and a special and amazing visitor from Sweden — the Bleriot XI.

It is hard to summarise exactly what makes Warbirds Over Wanaka the success that it is. My preceding text has listed some of the ground and aerial highlights that people from around the world have seen at Wanaka since 1988, but it hasn't answered the question adequately. The photographs in this book also illustrate the diverse nature of the aircraft, vintage and veteran machinery, fire engines and military vehicles that can be seen every two years.

I believe the answer lies in a combination of uniquely placed conditions that, when added together, provide the most cohesive, remarkable, non-stop action airshow and country fair in the world.

Were it not for the driving force and vision of Sir Tim Wallis there would have been no airshows, no videos, no books. Sir Tim has a special knack of gathering together fantastic teams of people; this can be seen in the other two members of his committee.

George Wallis has a single-mindedness about him, much like his brother. He has continued to expand the 'ground' side of the country fair — you only need to walk through some of the tractor and machinery displays at the peak of the mock attack to see the number of visitors that don't even glance skywards. Equally, the committee members of the Warhorses at Wanaka and Classic Fire Engine displays work tirelessly to provide fascinating displays.

The show's general manager, Gavin Johnston, has to be the best organiser I have met. It is his responsibility to promote, market, arrange, account and manage. Involved since 1990, he has undertaken his many tasks with total commitment — tasks that would require a team of four in many other committees!

When you add people like George and Gavin to the hundreds of others, both paid and volunteer, who work tirelessly to ensure a successful Easter weekend, you have most of the answer to the question of 'What makes Warbirds Over Wanaka a success?' People from all walks of life and all sections of the community pitch together to present the best warbird airshow in the world.

Ian Brodie
Curator
New Zealand Fighter Pilots Museum
December 2001

New meets old. A young girl meets an Ellis Chalmers tractor that would have been used by her great grandparents' generation! *Rob Neil*

Resting quietly, the pride of the 2000 show, Hurricane P3351, is put to bed for the night.

Dave Smith

A 1930s trailer pump, which would have been hooked up behind a fire truck or the fire chief's car.

Ken Chilton

1988 — The first show

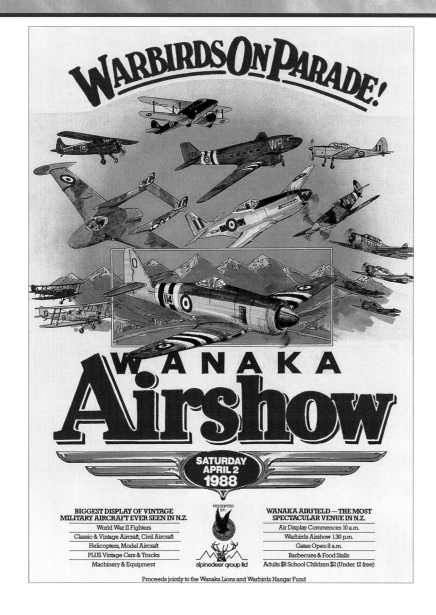

From an idea formed in 1987 and motivated by a wish to share his passion for aircraft, Sir Tim Wallis' Warbirds Over Wanaka has become one of the largest warbird airshows in the world. *Alpine Fighter Collection*

Pilot briefing, 1988. Tim can be seen issuing instructions second from right. *Alpine Fighter Collection*

First flown in February 1945, the Hawker Sea Fury is one of the fastest piston-engined fighters ever produced. This example was imported into New Zealand in 1986, and after being completely overhauled, flew again in March 1988. Her appearance at Warbirds On Parade was the highlight of the show for many. Her appearance also indicated the rapidly expanding interest and collecting that was occuring within the warbird scene in New Zealand. *Ian Brodie*

Imported by Sir Tim into New Zealand in 1984, the North American P-51D Mustang was a link back to his childhood days when he cycled to Wigram Air Force base to watch the type being flown by the Territorial Air Force. He immediately had his P-51 repainted to represent that flown out of Wigram by Squadron Leader Ray Archibald. *Ian Brodie*

Developed as a means of practising formation aerobatics, the New Zealand Warbirds' 'Roaring Forties' Harvard team has performed at many airshows around the country and in Australia, since 1987. Pictured are team members Steve Taylor (in aircraft) and Keith Skilling. *Ian Brodie*

Bottom left: The North American Harvard was used by the RNZAF as an advanced pilot trainer for over 30 years. At the time of their retirement in 1977, a number were purchased by New Zealand Warbirds members who syndicated the aircraft, allowing many members to return to a cockpit they had been very familiar with while serving in the RNZAF. *Ian Brodie*

Bottom right: John Denton doing a solo display at the 1988 show. *Otago Daily Times*

Left: The RNZAF have been enthusiastic performers at every show. In 1988 they attended with the Red Checkers aerobatic team as well as an Iroquois helicopter, a Strikemaster and a Friendship. Otago Daily Times

Right: World War I to scale. A 7/10th SE5A, flown by Tom Grant, leads two 7/10th Isaacs Furies flown by Rex Carswell and Dave Simpson, down the Wanaka airstrip. Otago Daily Times

Bottom left: The Tiger Top team. Rusty Butterworth atop a de Havilland Tiger Moth piloted by Tony Renouf. Otago Daily Times

Bottom right: John Denton displays the first civilian jet fighter in New Zealand. Owned by New Zealand Warbirds president Trevor Bland, the de Havilland Venom was flown by New Zealanders in both Singapore and Cyprus in the 1950s. This model was built in 1956 and was flown by the Swiss Air Force. Otago Daily Times

The words say it all. The location and southern hospitality were new to many who savoured the sun and enjoyed some amazing feats of flight and camaraderie between old and newly made friends. Otago Daily Times

Left: The formation aerobatics by Trevor Bland in the Mustang and John Denton in the Venom at the 1988 airshow were world class. Both retired fighter pilots from the RNZAF and RAF respectively, their slick display was indicative of their total trust in each other. Otago Daily Times

Right: One of the most recognisable aircraft ever made, the ubiquitous Douglas DC-3 first flew in 1935. This model was imported from Australia by a New Zealand Warbirds syndicate and provides a ready means of transport for members to airshows around the country. Ian Brodie

1990 — With a sneaky one in 1989!

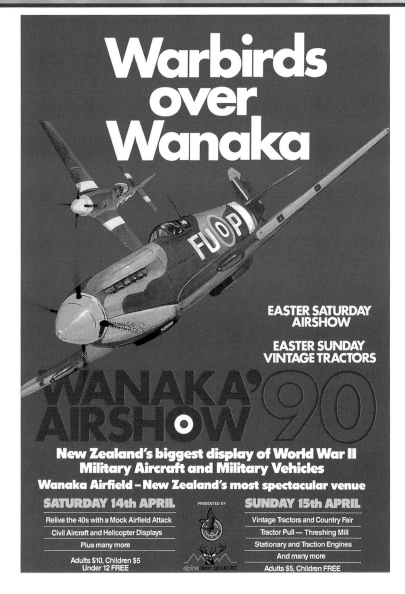

The Chance Vought F4U-1 Corsair was one of the greatest combat aircraft in World War II. First flown in 1940 it was the first single-engined US warplane to exceed 400 mph. Some 12,571 Corsairs were built, the last in 1952. Sir Tim acquired an early F4U-1 model from the USA in 1991. Here the newly arrived aircraft is being taken by barge from the port of Auckland to Hobsonville Airfield, prior to test flying. *Alpine Fighter Collection*

The debut of the Supermarine Mk XVI Spitfire in 1990 heralded the arrival of the aircraft that has become both an icon of aviation in New Zealand and the symbol of Wanaka. Built in December 1944 at Castle Bromwich in England, TB863 was utilised in the ground attack role by 453 Squadron (RAAF), in England. After many years languishing in storage she was rebuilt at Duxford (England), before arriving at Wanaka in 1989. *Phil Makanna*

Tim Wallis has flown over 300 hours in this aircraft, more than many World War II fighter pilots. Over 20,000 examples were built and there are now 50 airworthy around the world. *Phil Makanna*

Above: Tom Middleton eases the Spitfire towards a cameraship over Lake Wanaka. Many pilots have said that you wear the Spitfire like a glove, with every delicate movement becoming an extension of thought. *Phil Makanna*

Left: On a per capita basis more helicopters operate in New Zealand than in any other country. Here, Dennis Egerton provided a thrilling display in a Hughes 500D. *Otago Daily Times*

Right: Members of the RNZAF Skyhawk aerobatic team 'Kiwi Red' display their unique 'plugged formation' aerobatics. This was the year of the stealth fighter; Sir Tim and the team of commentators announced that a rare Stealth Fighter was inbound from Australia. As they talked to the pilot on his approach to Wanaka, all eyes turned to the left to catch a glimpse of the elusive fighter. He announced he had the airfield in his sight and as the 'invisible' aircraft dropped a bomb on the other side of the field an RNZAF Skyhawk roared in from the right. Low and very fast, the noise startled more than a few and created an onslaught of crying children. Otago Daily Times

Below: Where warbirds went to die? Rukuhia, near Hamilton, after World War II. Hundreds of ex-RNZAF aircraft were stored here before being sold and melted down for pots and pans. Sir Tim has always lamented the loss of so many of these famous aircraft. By 1990 his plans to bring some of these classic fighters back into the New Zealand psyche were reaching fruition. Bryan Cox

The New Zealand Warbirds Association's Roaring Forties team practising over Lake Wanaka. *Ian Brodie*

1992 — The year of the Me-109

Pilots of the AFC aircraft: (from left to right) Tom Middleton, Rex Dovey, John Lamont, Keith Skilling, Phil Murray, Simon Spencer-Bower and Sir Tim Wallis. *Ian Brodie*

The 1992 airshow will be remembered by many as 'The Year of the 109'. Looking for exciting acts to add to the show, the committee discussed options with Ray and Mark Hanna of the UK-based Old Flying Machine Company. It was decided to import a Messerschmitt Me-109J for the show. To help sponsor the costs a Messerschmitt Gold Pass area was erected, with visitors paying a premium for grandstand seating set up in a prime position. *Phil Makanna*

The publicity surrounding the Me-109 ensured that a crowd of over 50,000 attended the 1992 show. The opportunity to see two of the most famous aircraft in the world in a dogfight was not to be missed!

Ray (in the Spitfire) and Mark (in the 109) provided an impeccable display and to this day many people consider this the highlight of all Warbirds Over Wanaka shows.

Ironically, this Me-109 was fitted with a Rolls Royce Merlin engine — the same as the Spitfire. The aircraft was built in 1943 in Germany and shipped to Spain minus her engine. It was eventually fitted with a Merlin and given the name HA-1112 M1L (Buchon).

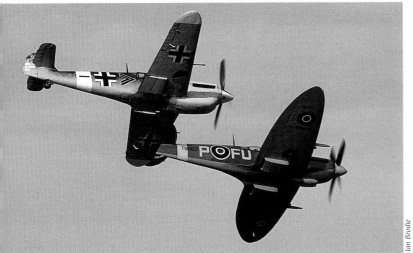

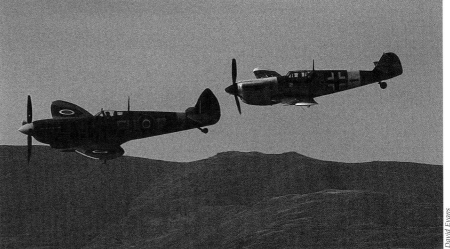

After service with the Spanish Air Force, this Me-109 was one of a number purchased in 1966 for the film *The Battle of Britain*.

Purchased by an American after the aircraft's film commitments were fulfilled, she returned to the UK in the late 1980s and was bought, after an extensive rebuild, by the Old Flying Machine Company.

Soon to become a warbird in its own right — the BAC Strikemaster makes its last appearance at Wanaka before retirement from RNZAF service. Utilised as an advanced jet trainer, the type entered service in 1972 as replacement for the venerable Harvard. *Rob Neil*

How nice of you to drop in! Members of the Southland Parachute Club and Otago Skydiving Club provided spectators with a freefall parachute display. Otago Daily Times

Sir Tim Wallis is acknowledged as an expert at flying helicopters and has over 12,000 hours on many types. As well as undertaking the demands of being chairman of the airshow, he took time out on the Saturday morning to pair with another expert, Dennis Egerton, to provide a synchronised Hughes 500D display. *Barry Harcourt*

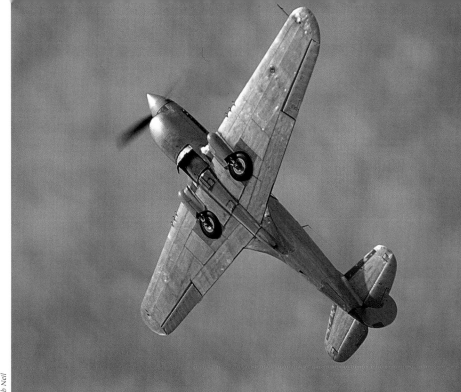

One of Tim's desires was to see the return of the two main fighters flown by the RNZAF in the Pacific during World War II. In 1992 he achieved this with the debut of both the Curtiss Kittyhawk and the Vought Corsair. Purchased by Sir Tim in 1988, the Kittyhawk was rebuilt at Wanaka by a team of engineers. As the airshow deadline crept closer, Chief Engineer Ray Mulqueen had a simple sign on the door outside his office — 'When It's Ready'! After many thousands of hours of work the first engine run was undertaken on the Thursday prior to Easter.

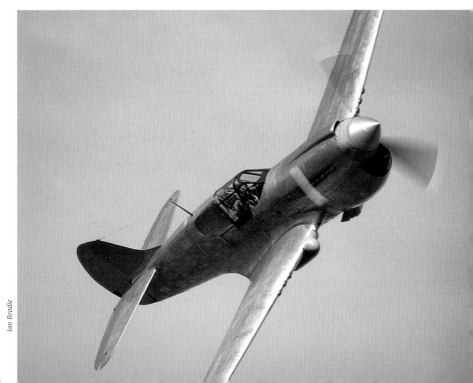

After a four-year rebuild, the Curtiss P-40K Kittyhawk made its maiden flight in the hands of Mark Hanna on the Saturday morning of the show. She is in bare metal as there was no time left to apply the paint scheme! Otago Daily Times

A distinctive design characteristic of the Vought Corsair is the gull wing that allowed enough clearance for the huge propeller as well as giving more stability in a dive. A total of 424 examples were operated in the ground attack role by the RNZAF in the Pacific during World War II. Most of the Corsairs flown back to New Zealand at the cessation of hostilities ended their lives at Rukuhia (near Hamilton) as scrap. Post-war they were operated by 18 Squadron as part of the occupation force in Japan. Prior to the unit returning home all aircraft were destroyed in a huge bonfire. There is currently only one airworthy RNZAF example, in the UK, owned by Ray Hanna.

Above: Commentators' position — Kiwi style! *David Evans*

Left: The Grumman Ag-Cat operated by Phil Maguire of Queenstown-based Pionair Adventures. Although designed as a crop-duster, the type proved very suitable as a passenger aircraft — two barnstorming tourists sitting side by side on seats that replaced the original hopper. *Rob Neil*

Below: Operated by the RAF in the 1930s, the Hawker Fury was one of the most stylish biplanes produced, as testified by this 7/10th scale Isaacs Fury II. First flown in 1982, the aircraft was displayed by Aucklander Rex Carswell. *Alpine Fighter Collection*

Resplendent in red — the New Zealand Warbirds' de Havilland Canada Beaver at the hands of Peter Rhodes. A number of these aircraft were operated in a top-dressing role in New Zealand. However, the red scheme represents NZ6001, which was part of the Trans-Antarctic Expedition in the 1950s. *David Evans*

The Cessna L-19 Bird Dog was developed as an observation aircraft, and the type was used to great effect as a Forward Air Control aircraft during the Vietnam War. This example served with the US Army and South Vietnamese Air Force before being shipped to New Zealand in 1990. *Ian Brodie*

Designed in 1944, the de Havilland Dove was one of the first new generation post-war British commercial aircraft. Known as the Devon in military service, the RNZAF operated 30 examples before retiring them in 1981. *Ian Brodie*

 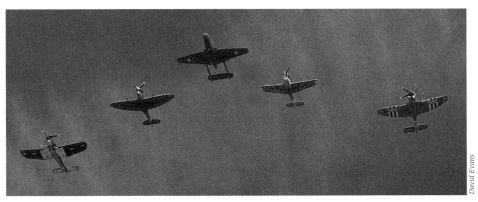

A formation that may never be repeated in New Zealand. The Corsair, Spitfire, Vampire, Me-109 and Sea Fury provided an exciting combination as the final display to Warbirds Over Wanaka 1992 — all to the stirring music of Vera Lynn.

The RNZAF Kiwi Blue Parachute Team led by Captain John Tinsley launched the lunchtime Air Force display slot. *David Evans*

Smoke swirls around a 'bad-guy' Harvard during the mock attack at the end of the show. Always one of the most popular sequences, the skies of the Upper Clutha River come alive with aircraft accompanied by very realistic explosions on the ground. *David Evans*

1994 — The 'sinking' of the *Yamato*

During production the Spitfire changed in many ways, including a doubling in horsepower. Sir Tim was keen to have three different examples and in 1994 the second arrived at Wanaka in the form of a Mk XIV powered by a 2035 hp Rolls Royce Griffon engine. Mark Hanna flew the aircraft from Auckland to Wanaka on Good Friday.
Ian Brodie

Sir Tim flew the Mk XIV for the first time later that year and compared her to a wild gypsy woman after flying the more sedate Mk XVI.
Ian Brodie

Always looking for special acts, in 1994 the committee presented an amazing two-wheeled machine that never left the ground. John Britten showcased his revolutionary Cardinal Britten V-1000 motorcycle in a race against a helicopter and a Pitts Special. At the hands of Loren Poole the motorcycle won by a country mile and amazed the crowd with its acceleration and speed. *David Wethey*

Another New Zealand first was the helicopter drop of a glider. Both glider pilot (Bruce Drake) and helicopter pilot (Alan Bond) described the event as 'interesting'. *Alistair Kinniburgh*

Left: Two newcomers to the airshow in 1994 were the Mitsubishi Zero replica and Grumman Avenger. In this scene the Zero attacks the Avenger during the Pacific theatre re-enactment. The Avenger is shot down, but the Zero is then despatched by a marauding P-40 Kittyhawk. *Ian Brodie*

Sir Tim loved flying his Spitfire around the Southern Alps and was always keen to get photos of her in this mountainous environment. Here, Simon Spencer-Bower and Ian Brodie await the arrival of the Spitfire. There is nothing more amazing than standing 3000 metres up on the side of a mountain listening to a Merlin engine.

Above left and above: The second de Havilland Venom to fly at Wanaka was also imported from Switzerland. Designated an FB1 Mk 4, it is powered by a 4850-pound Ghost 48 Turbojet, and is seen above left employing the once common cartridge start. Piloted at the show by Trevor Bland, her spectacular colour scheme is that of the solo display aircraft of the Swiss Aerobatic team — Patrouille de Suisse.

Left: The Imperial Japanese Navy battleship *Yamato* floats again — well, almost. From an idea by Sir Tim to fruition in four days, the 200-foot canvas profile underwent ceaseless attacks from the Corsair, Kittyhawk and Avenger. *Mike Provost*

Far left: More and more aircraft were making their way to Wanaka from north Russia. On the hangar floor in the New Zealand Fighter Pilots Museum were the remains of two Junkers Ju-87 Stuka and a Hawker Hurricane. *Alistair Kinniburgh*

Left: Sir Tim had also acquired two Messerschmitt Bf-110 twin-engined fighters. One had spent 50 years in a grave of fresh water before being discovered, surprisingly intact. The wrecks were subsequently sold to the Techniks Museum in Berlin, where one has now been restored to static display status. *Alpine Fighter Collection*

Built in 1952, this Sea Fury was operated by the Iraqi Air Force for a number of years. Her colour scheme is that used by Lieutenant 'Hoagy' Carmichael of the Fleet Air Arm during the Korean War. Lt Carmichael was responsible for claiming the Royal Navy's first jet victim — a MiG 15.

Alpine Fighter Collection

Toby Clark amazed the crowd when he flew this BK 117 into amazing attitudes and manoeuvres.

Alpine Fighter Collection

Phil Makanna

A number of famous World War II fighter pilots were invited to attend the show and help celebrate the fiftieth anniversary of the D-Day landings at Normandy. From left to right: (back row) Ian Brodie (Curator, NZFPM), Des Scott, Johnnie Houlton, Jamie Jameson, Johnny Checketts, John Patterson; (front row) Sir Tim, Johnnie Johnson and Jim Sheddan. *Mike Provost*

Above left: In 1994 Mo McAuliffe was re-acquainted with TB863, the aircraft he had formerly flown in early 1945 as a member of 453 Squadron. *Mike Provost*

Above right: Air Vice Marshal 'Johnnie' Johnson CB, CBE, DSO and 2 bars, DFC and bar, in the Mk XIV Spitfire. Johnnie survived the war to become the RAF's top-scoring ace, with 38 enemy aircraft to his record. Sadly, Johnnie died in 2000. *Phil Makanna*

The business-end of the Mk XIV Spitfire. *John Dibbs*

Richard Hood launches his Pitts Special (with appropriate registration). Designed in 1946, this aircraft is fully aerobatic. *Phil Makanna*

A newcomer to Wanaka, the Cessna A-37 Dragonfly was used by the USAF as a primary trainer. This particular version saw service during the Vietnam War in the ground attack role. Manufactured in 1972, the aircraft had only just flown prior to the airshow after a complete rebuild — hence the lack of paint scheme.
Phil Makanna

One of the largest single-engined aeroplanes ever built, the Grumman Avenger was designed as a carrier-borne torpedo bomber, hence the folding wings. This example made her debut in an American colour scheme, after being purchased by Sir Tim from the Old Flying Machine Company.
Ian Brodie

Simon Spencer-Bower (left) and Tom Middleton after Simon's first flight in the Mk XVI Spitfire. Tragically, Tom was killed in an aircraft accident in December 2000. *Ian Brodie*

Ian Reynolds and Simon Spencer-Bower in their de Havilland Chipmunks. *Mike Provost*

Considerable effort is made by a number of pilots who fly their biplanes to Wanaka from around New Zealand. None of this 60-minute flight, air-conditioned jet business! Here a group of Tiger Moths prepare for their formation flypast over a North American Harvard. *Mike Provost*

The scourge of the Western Front. In 1917 the Fokker Triplane ruled the skies and shot down many times their own number at the hands of skilled German pilots, including Baron Manfred von Richtofen. Stuart Tantrum built this full-size replica with John Lanham in the 1980s. *Otago Daily Times*

Designed in 1932 as a light transport aircraft, the de Havilland DH-83 Fox Moth has many links with New Zealand. Flown on the West Coast of the South Island by Air Travel, this model was originally owned by HRH The Prince of Wales as part of the Royal Flight. Tim's father flew in her many times along the West Coast while attending to his timber business. Rebuilt by Colin Smith of the Croydon Aircraft Company at Mandeville, she wears her royal plumage.
Alpine Fighter Collection

The 1994 Red Checkers team, led by Squadron Leader Ian McClelland. Flying the 210 hp CT-4B Airtrainer, the team performed some very exciting manoeuvres, including the famous 'mirror'. *Mike Provost*

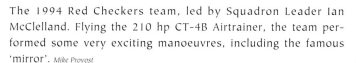

In or out? An RNZAF Iroquois shows its ability to winch a stranded crew member aboard while Mark Frood performs a spectacular bungee jump from a Helicopter Line Squirrel.

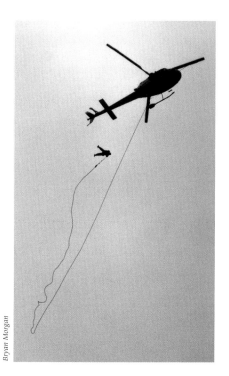

The large glass canopy of the Harvard provides an excellent view for both instructor and student, as can be seen in this picture taken from the back seat while flying over Lake Wanaka.
Ian Brodie

Above left: Friend and foe meet above the Upper Clutha River. The Corsair in formation with the Mitsubishi Zero replica purchased by Sir Tim in 1993. Converted from a North American Harvard for the film *Tora, Tora, Tora* in the 1960s, the aircraft was purchased from the Old Flying Machine Company. *Ian Brodie*

Left: In 1984 this scene would have been bar talk. Not one, but two Spitfires and Mustangs with a Corsair and a Kittyhawk provided the fighter finale for Warbirds Over Wanaka 1994.
Bryan Morgan

Sir Tim in his beloved Spitfire with John Lamont in the P-40 Kittyhawk and Tom Middleton in the Chance Vought Corsair. *Ian Brodie*

1996 — Tim's absence

Just a bit too late to be part of the 1996 airshow, one of the first Polikarpov I-16s to arrive in Wanaka receives a good old Kiwi welcome following its first New Zealand flight from staff and families, a toast of vodka. From left to right: (forward of wing) Malcolm Brown, Greg Parker, Sir Tim, Ray Mulqueen, Ewan Fallow, Keith Skilling; (aft of wing) Grant Bisset, Dianne Brodie, Ian Brodie, Jonathon Skogstad, Annie Trengrove, Gavin Johnston, Angie Brown, Chrissy Fallow, Sharlene Mulqueen and Kaz. *Prue Wallis*

Warbirds Over Wanaka would not be possible without a dedicated team of volunteers. Stan Vowles (left) checks the parking situation with Rex Whiteside. *Mike Provost*

Above and above right: Buoyed with the success of the visiting Me-109 in 1992, the committee decided to bring to New Zealand a very rare Messerschmitt Me-109 G10, powered by an original Daimler Benz engine. Eased from its cocoon, the aircraft was re-assembled by Roger Sheppard from the Old Flying Machine Company assisted by Alpine engineers — all under the watchful gaze of Mark Hanna.

Right and above far right: The sound of the Daimler Benz engine with supercharger screaming was amazing. The Me-109, one of only two airworthy examples in the world, performed both a solo routine and a mock dogfight with the newly arrived AFC's Yak-3M.

Over 35,000 Me-109 fighters were produced between 1936 and 1945. Sadly, few now remain. It was a poignant moment for many as a group of Spitfire and Tempest pilots gathered to meet with a former adversary. Pictured are: (left to right) Max Collett and Maurice Mayston (485 Squadron), Jim Sheddan, Jim McCaw and Jack Stafford (486 Squadron), Ray Hanna (Old Flying Machine Company) and Bill Miller (486 Squadron). Jack Stafford shot down one of these types flying a Hawker Tempest over Germany in 1945.

Making her airshow debut, the Yakovlev Yak-3M amazed many with her manoeuvrability. Despite over 37,000 fighters being produced by the Yak family, very few existed by the 1960s. In the 1980s Sergei Yakovlev (son of the original designer) re-opened the original factory and produced a small number of new models. Externally identical, the only change was the fitting of an American Allison engine.

In an effort to stimulate the growth of aviation art in New Zealand, the New Zealand Fighter Pilots Museum staged an art exhibition that attracted the largest number of original aviation paintings ever assembled in New Zealand. Wellington artist Ron Fulstow is with two of his own works. Ron is a superb artist and his paintings hang in galleries and museums around the world. *Ken Chilton*

Race time! Harvey Hutton, in the Hughes 500, pitted himself against Grant Bisset in a Pitts Special, a 1971 March racing car and a Western Star 592Z Superstar truck. The race was a dead heat, possibly as a result of the heavy cost involved in losers' bets.

Alpine Fighter Collection

Another Messerschmitt appeared at Wanaka in the guise of an Me-108 Taifun. Owned by Colin Henderson and Maurice Hayes of Auckland, the type was first flown in 1934 and used extensively by the Luftwaffe in the liaison role.

Alpine Fighter Collection

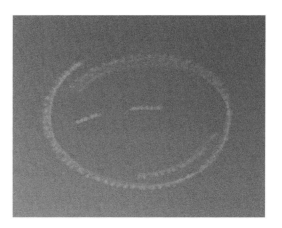

George Wallis in the cab of one of his double-decker buses. *Alpine Fighter Collection*

Robert Koch expertly flew his gyrocopter high in the Wanaka sky. *Alpine Fighter Collection*

Former UK aerobatics champion Mark Jefferies provided the crowd with a dazzling array of stunts in his Yak-50. To close the Sunday afternoon display he 'smiled' on everyone from aloft (see above right). *Ian Brodie*

Sir Tim was not present atop the commentary bus this year. Seriously injured in a take-off accident in his Mk XIV Spitfire in early January, he watched the show from a nearby hill. Ray Hanna took over the Mk XIV duties. *Phil Makanna*

Above: With seating for eight passengers, the de Havilland DH 89 Dragon Rapide first flew in March 1937. A number made their way to New Zealand and during World War II, known as the Dominie, they operated with the RNZAF before returning to their civilian guise. AKY is now a resident in Queenstown, where she allows passengers to experience flight 1930s style. *Dave Miller*

Right: Other affordable warbirds appearing in New Zealand in increasing numbers are the Yak-52 and the Nanchang CJ-6. Economical to run and entirely aerobatic, they are often seen in flocks as their pilot/owners converge on aerodromes to good-naturedly argue which is the better. At right, Sir Kenneth Hayr is seen in his Yak taxiing for take-off. Sir Kenneth was tragically killed in England on 3 June 2001. *Geoff Sloan*

A couple on a globe-trotting tour also made their way to Wanaka in 1996. John and Joyce Proctor arrived before Easter in their immaculately restored Grumman HU-16 Albatross, which they rescued from a junkyard after it was retired from the US Navy. After five years' work their flying/floating palace was ready for them to enjoy. They set off on a journey that took them from the USA to Russia and then to New Zealand.
Geoff Sloan

Another larger amphibian that is a popular visitor to Wanaka is the Consolidated PBY Catalina. Characterised by its huge wing area, the type was used by a number of countries during World War II in both the anti-submarine and air/sea rescue role. Ferried from Africa in late 1994, she is an important addition to the New Zealand warbird scene representing the 56 examples operated by the RNZAF during and after World War II.
Mike Provost

The Beech D17S Staggerwing epitomises the art deco style of the 1930s. First appearing in 1932, it features a retractable undercarriage and rich leather upholstery, suitable for the nouveau riche of the time. Imported into New Zealand in 1995 by Robin and Elaine Campbell, they show justifiable enjoyment in displaying their elegant aeroplane. *Geoff Sloan*

Basic trainer for most American pilots during World War II, the Boeing Stearman is powered by a 220 hp Continental W-670 radial engine, which gives it more power than its Commonwealth counterpart, the Tiger Moth. Owned by Dave Horsborough and based in Christchurch, this distinctive blue biplane takes passengers for open-cockpit joyrides. *Mike Provost*

As the airshow increases in stature and size, meticulous pilot briefings have become essential. *Shar Devine*

Asked to design a replacement for the venerable Harvard, North American designers produced the two-seat T-28 Trojan in 1949. Utilised by the United States Air Force and United States Navy, this model was imported into New Zealand by the late John Greenstreet in 1989 after languishing in the dry desert sun of Arizona. It wears the colours of VA122 of the Pacific Training Fleet based at Leemore, California. *Alpine Fighter Collection*

The Westland Scout helicopter (top) was utilised by the British Army for many years and the Bell OH-58A Kiowa by the US Army. *Mike Provost*

Above: The largest single-engined biplane ever produced, this distinctive Antonov AN-2 Colt was flown from Lithuania to New Zealand by her owner, Neville Cameron. *Paul Hillier*

Right: Phil Murray provided some light relief in the Piper Cub. His 'crazy' flying technique made it look like the aeroplane would crash at any moment. However, Phil is a very experienced pilot and the 'craziness' belies the skill involved. *Mike Provost*

Resplendent in her RNZAF 15 Squadron colour scheme, the AFC's Kittyhawk is flown here by John Lamont. John spent twelve years in the RNZAF and is a very experienced warbird pilot. He is the Flying Co-ordinator for Warbirds Over Wanaka. *Phil Makanna*

The bulk of an RNZAF Lockheed C-130 Hercules dwarfs the taxiing CT-4 Airtrainers of the Red Checkers. The 'Herky-Bird' has been in service with the RNZAF since 1965. *Geoff Sloan*

The rarest aircraft to be displayed in 1996 was the AFC's Nakajima Ki-43-1c Hayabusa. Purchased by Sir Tim from Australian collector Col Pay, this was the only complete Type 1 in the world. Restored to taxiable condition at Wanaka, this aircraft was flown by a number of Japanese Sentai based at Rabaul, New Britain. Here Simon Spencer-Bower displays typical impudence to his adversary, the **Corsair**. *Mike Provost*

Far right: The corporate area of the airshow allows businesses from around New Zealand to entertain guests in style with the best view in the house. *Shar Devine*

The Confederate Air Force in New Zealand operate this Beech 18 Utility Transport from their Dairy Flat headquarters in Auckland. Although designed in the 1930s, the type is still in use in various countries around the world. *Phil Makanna*

The colour scheme of the Warbirds' Dakota is that of an aircraft flown by New Zealander Rex Daniels when he dropped parachutists over the Orne Canal on D-Day, 6 June 1944. Post-war, Rex was one of the founders of South Pacific Airlines of New Zealand during the 1960s.
Alpine Fighter Collection

Veteran road-safety campaigner Max Corkill and his travelling companion, Rastus. Sadly, both were killed in a road accident in January 1998.

Above: With almost every underwing pylon full, Brian Rhodes lifts off in the Cessna A-37 Dragonfly. The type is equipped with 12 underwing hard-points and a mini-gun for use in the COIN (counter-insurgency) role. *Mike Provost*

Below: Tom Middleton had a special love-affair with the Corsair. During World War II his father John flew the same type with the Fleet Air Arm and undertook a number of missions over Japan. This Corsair is an early 'bird-cage' model (which refers to the cockpit canopy) and spent her wartime career as a trainer, amassing 776 hours in that role. *Ian Brodie*

Left: The Aviation Trade Expo at Wanaka has become a key venue for many exhibitors of aviation-related goods. Here, the 'big top' is erected to house many of the displays. *Shar Devine*

Above: The Tiger Moth formation marches gently across the sky in the morning and is replaced by the Fighter Finale in the afternoon. It is doubtful that this line-up has been seen anywhere else in the world: an Me-109, Spitfire, Yak-3, Corsair, P-40 and Mustangs.

Left: The Alpine fighter trio of Spitfire, Yak-3 and Mustang. *Mike Provost*

Diversity on the ground as well as in the air. The names Burrell, Rumsey and Dennis are just as familiar to the many enthusiasts at Wanaka who prefer to keep their eyes firmly on the ground.

1998 — Red Stars Rising

The wonderful combination of people and aeroplanes was again prominent in 1998. With Sir Tim is special guest Brigadier General Charles (Chuck) Yeager (right), who was the first man to break the speed of sound. In the air the Polikarpov (below) reigned.

It was indeed Red Stars Rising as engineers and staff involved in the Polikarpov rebuilds travelled from Russia to witness an historic event. Pilot Steve Taylor displayed what 1000 hp means, both in the air and on the ground.

Phil Makanna

Prue Wallis

Phil Makanna

'Nasdarovia' to the I-16 — newly rated pilots Keith Skilling, Mark Hanna, John Lamont and Steve Taylor toast the delights of these small and chubby but aggressive fighters. From a plan initiated by Sir Tim in 1990, the aircraft were rebuilt from wrecks recovered in Russia. Using original plans, six of these aircraft were completed over the ensuing six years.

Phil Makanna

Ian Brodie

Phil Makanna

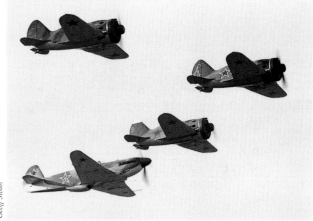

It would have been over 50 years since anybody had seen this number of I-16s in the air at once and they stunned the crowd with their amazing noise and manoeuvrability. The pilots experience a blast of wind in their faces and a loud roar in their ears. Simplicity in design was the key factor for this aircraft. For example, the undercarriage is operated by a cable attached to a handwinch in the cockpit and there is no trim or flaps. This sparse design undoubtedly contributed to the success of the I-16 in the harsh Russian conditions.

Adding to the Russian theme, dancer Masha Volobueva provided graceful entertainment.

Above: A record number of entries were received for the Aviation Art Exhibition in the New Zealand Fighter Pilots Museum, including this piece dedicated to New Zealand aviation pioneer, Richard Pearse.

There is only one recorded New Zealand fighter pilot who flew with the United States Army Air Force during World War II. Flight Lieutenant Jack Cleland DFC spent six months in 1944 on secondment to the 357th Fighter Group. Flying his P-51D Mustang 'Isabel', which was named after his wife, one of his squadron compatriots was Chuck Yeager. In Jack's honour the AFC's Mustang was painted in his colour scheme. Although Jack died some years ago, his wife Isabel (above) and family came to Wanaka. Here she met her namesake for the first time and enjoyed a ride in the aeroplane after the show.

Phil Makanna

New Zealand Warbirds Association President Trevor Bland. Trevor retired from an Air Force career to fly with Air New Zealand and was the pilot of Tim's first Mustang for a number of years. *Geoff Sloan*

The very enthusiastic Warhorses at Wanaka team organises and arranges a huge cadre of military vehicles and equipment to visit Wanaka every two years. *Rob Neil*

The name 'Red Checkers' became a misnomer in 1998 when the RNZAF team displayed their new, highly visible yellow colour scheme. *John Dibbs*

The sheer miscellany of aircraft that visit Wanaka during the airshow is no more apparent than in this photo. A de Havilland Vampire taxies past a P-51D, a Fouga Magister and a Cherokee. *Geoff Sloan*

A superb de Havilland study. From left to right: a DH-82 Tiger Moth, DH-84 Dragon, DH-89 Dominie and a DH-90 Dragonfly bask in the sun. *Geoff Sloan*

Back flying over her home turf, this Mandeville-based de Havilland DH-83 Fox Moth was one of New Zealand's first commercial airliners. Based with Air Travel on the West Coast of the South Island for many years, she was immaculately restored by Colin Smith and the Croydon Aircraft Company. *John King*

Rex Dovey has been part of the Central Otago flying scene for many years. Flying helicopters for Sir Tim during the period of early deer recovery operations, he is now based in Queenstown, flying tourists to Milford Sound. Formerly the primary pilot of the AFC's Grumman Avenger, Rex now flies the Polikarpov types I-16 and I-153. *Shar Devine*

Australian Nigel Arnot provided an amazing display in his Sukhoi Su-31. The nine-cylinder 400 hp aircraft allows for fantastic aerobatics — including the flyaway tailslide, a loop from below stalling speed and the 'Harrier Pass', where Nigel hovers the aircraft.

Alpine Fighter Collection

Rob Neil

Geoff Sloan

Making her first appearance at Wanaka was this Curtiss P-40E Kittyhawk (NZ3009), owned by the Old Flying Machine Company. The only airworthy example of 297 of these aircraft operated by the RNZAF, she looked resplendent as the Wairarapa Wildcat — the aircraft flown by top-scoring Commonwealth fighter ace in the Pacific, Geoff Fisken. *Phil Makanna*

Above: When the name warbird is mentioned, most people tend to think this means piston-engined aircraft. However, a number of jets are now appearing in New Zealand and the MiG-15 is no exception. First flown in 1947, the MiG-15 became the scourge of the Korean War, possessing a better rate of climb, turning circle and service ceiling than the American F-86 Sabre. This two-seat trainer was operated by the Polish Air Force until 1989. *John Dibbs*

Top right: Airshow cuisine of the best degree as prepared by Wanaka Lions Club members. Food stalls are operated by local community organisations, with all profits returning to them for use in the district. *Geoff Sloan*

Middle right: Blenheim-based Steve Petersen with his Nanchang CJ-6, China Doll. *Geoff Sloan*

Right: Jerry Mead is one of the most accomplished airshow commentators in the world. Invited to Wanaka he dispensed his knowledge and brand of humour to the crowd in 1996 and 1998. *Phil Makanna*

Making her last appearance at Wanaka before retirement, an RNZAF Hawker Siddeley Andover makes a short-field landing. Operated by the New Zealanders since 1976, the type was used for short- to medium-range tactical transport. *John Dibbs*

Far left: 'Steady as she goes, Mainwaring!' Warhorse members undertake a reconnaissance jaunt in preparation for their 'Dad's Army' routine. *Geoff Sloan*

Left: The boys who make the toys tick. Some of the AFC's engineers. From left to right: Matt Bailey, Greg Parker, Malcolm Brown and Ewan Fallow. *Ian Brodie*

You can even post your mail at the show using specially designed stamps. *John Dibbs*

Above and below, left and right: The sights and, in particular, the sounds of formation flypasts are always a crowd-pleaser. Mustangs are accompanied by a Corsair, a Kittyhawk and a Spitfire, as well as the ubiquitous Harvards, eleven-strong. *Rob Neil*

2000 — The Hurricane completed

It was a special moment when Hawker Hurricane P3351, owned by the AFC, was completed in early 2000. The airframe was reconstructed by Hawker Restorations in the United Kingdom before being shipped to New Zealand. Air New Zealand Engineering Services staff in Christchurch then spent the next six years returning the aeroplane to better than original condition. Below, Sir Tim with Air New Zealand staff (left to right) Kevin Nicholls, Graeme Wilson and Ian Carmichael. AFC pilot Keith Skilling flew the Hurricane on her maiden flight on 12 January 2000.

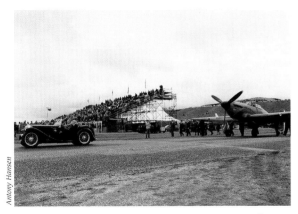

Two New Zealanders are known to have flown Hurricane P3351. Coincidentally Ness Polson and William (Dusty) Miller were both born in Invercargill and flew her in England while with 55 Operational Training Unit in 1941. An even more amazing fluke was that Dusty Miller retired to Wanaka in 1978. After the opening of the New Zealand Fighter Pilots Museum he became a volunteer guide. In a very moving tribute to all who flew the aeroplane and to Sir Tim and the dedicated team who rebuilt her, Dusty Miller was escorted by Warhorse members and Air New Zealand staff to the front of the Gold Pass area.

Dusty Miller answering an onslaught of media questions from 'his' aeroplane after she arrived from Christchurch. Sadly, Dusty passed away during 2001.

Built in early 1940, P3351 survived three major battles and is now one of only ten airworthy examples in the world. After seeing action in the Battle of France and Battle of Britain, she was used as a trainer. She was shot down in Russia by ground fire in the summer of 1943 and spent the next 50 years in the tundra.
Ian Brodie

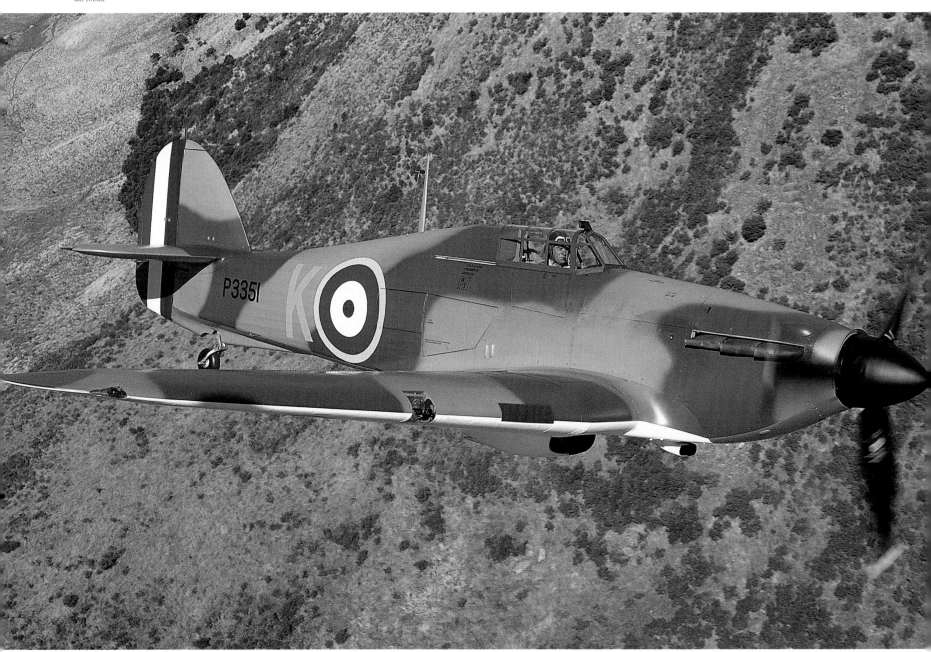

Paired with the more famous Spitfire, the two aircraft provided another New Zealand first for airshow visitors.

The Hurricane appears in the scheme worn while with 73 Squadron in France, including the unique black-and-white undersurfaces. From 3 to 18 June 1940, P3351 took part in a battle that was doomed to failure. She was one of the last aircraft to leave France after the German invasion.

Sir Tim watches as Keith Skilling prepares to commence his display in Hurricane P3351. *Geoff Sloan*

Pilot Keith Skilling is given a lift to P3351 on the same type of vehicle that would have been used during World War II, a BSA 350 cc army dispatch motorbike. *Chris Hinch*

Local volunteer fire brigades have provided important services to each show. It is the support of these local community groups that are an essential part of Warbirds Over Wanaka. *Alpine Fighter Collection*

Dusty Miller with Sir Tim after the Hurricane's display. *Prue Wallis*

A 1942 M3A1 Scout car leads a Warhorses over Wanaka team along the display line. *Phil Makanna*

Trevor Bland lands the Fouga Magister resplendent in the colour scheme of the French Aerobatic Team, the Patrouille de France. This Magister was imported to New Zealand by the late Dougal Dallison. *Ken Chilton*

Steve Taylor puts his Edge 540 through its paces. With wings made entirely of composite materials, which have been statically tested to ± 23G, the aircraft is one of the most advanced purpose-built aerobatic aircraft in the world. *Rob Neil*

Our great friend, Tom Middleton, who died on 15 December 2000. *Phil Makanna*

News of another P-51D Mustang fresh from rebuild and due in Wanaka on the Sunday morning was surreptitiously filtered to the crowd. All eyes looked to the runway as the aircraft landed and then the penny dropped — who had ever seen an aircraft with the codes WOW and serial 2002? *Prue Wallis*

Major sponsor of the show Air BP ensures there is plenty of gas for all the aircraft over the weekend. Malcolm Brown is ably assisted by Wanaka schoolboy volunteer Tom Paget. *Chris Hinch*

Sir Tim and his youngest son Nick are escorted onto the flightline for the Hurricane display. *Chris Hinch*

The Nakajima Hayabusa made her last appearance at Wanaka in 2000. Now resident in the USA, her new owner is undecided about whether to rebuild her to flying condition. *Otago Daily Times*

Transported back to the dawn of flight, the Bleriot XI was a huge crowd-pleaser. Imported from Sweden for the show by Mikael and Gunilla Carlsson, this is one of only three airworthy in the world. Built in 1918 and rebuilt in the early 1990s, the Bleriot fascinated everybody, especially the other pilots in attendance.

In 1909, the Bleriot XI was the first aircraft to cross the English Channel. It was also destined to become a New Zealand aviation pioneer. In 1913 American Arthur 'Wizard' Stone imported an example and undertook the first long sustained and truly practical flight in New Zealand at Auckland's Alexandra Park. The Bleriot is powered by a 50 hp Gnome Omega Rotary engine and cruises at 42 knots. In 1999 Mikael Carlsson recreated history when he flew the aircraft across the English Channel.

A trio never seen together in the air before. Another world first in 2000 was the debut of the Polikarpov I-153. Here, one teams up with the Hurricane and Spitfire for this special photograph.
Phil Makanna

The show's over and now we can celebrate! From left to right: John Peterson, Tom Middleton, Boris Osetinskiy (the AFC's Russian partner), Sir Tim Wallis, Keith Skilling, Steve Taylor, John Lamont, John Lanham and Stu Goldspink. *Prue Wallis*

Another newcomer to Wanaka was the Kaman Seasprite, which is operated by the Naval Support Flight of No. 3 Squadron RNZAF. Replacement to the ageing Wasp, the helicopter features a retractable undercarriage and is capable of speeds of up to 150 knots. *Rob Neil*

Right and top: The Fletcher topdresser as a formation aerobatic aircraft? Only in New Zealand! Wanganui Aerowork has maintained an interest in formation and display flying over most of its 50 year history. Utilising the Fletcher and its progeny, the Cresco, the team performs a stunning display of low-level formations, opposing pair displays and smoke-filled antics.

Rob Neil

Top left: Pick your nose — the Sunday afternoon line-up provides a wonderful assortment of aeroplanes. *Ken Chilton*

Far Left: Graeme Taylor is the organiser of the static engine displays. He can also be seen in the air, the proud owner of a Kitfox homebuilt aeroplane. *Rob Neil*

Left: Polikarpov I-153 pilot John Lanham has been involved in aviation all his life. After spending 26 years in the RNZAF where he was CO of 14 and 75 Squadrons (which flew Skyhawks), he is now general manager of General Aviation Operations for the Civil Aviation Authority. *Chris Hinch*

Above: First flown in 1938, the Polikarpov I-153 was based on a restressed version of the I-152 biplane, but with the upper gull wing of the I-15. Powered by a 1000 hp radial engine, three of these examples were rebuilt in Russia for Sir Tim over a six-year period. They are the only airworthy examples in the world.

Ken Chilton

This is a sight that will probably never be repeated. Five I-16s perform a ground attack mission before teaming up with three I-153s to provide a formation flypast. Since 2000, two I-16s and an I-153 have been exported to the US for future sale. One each of the types will be retained at Wanaka.

Dave Smith

Phil Makanna

Dave Smith

For the first time the entire sound system for the airshow was controlled though computers supplied by Compaq and Computerland.

Chris Hinch

Above: Although most of the 2000 Easter weekend was dogged with untypical gloomy weather, blue sky was still found to promote the major sponsor. *Phil Makanna*

Phil Makanna

The ability to quickly apply a new colour scheme means aircraft can be painted in colours used in different conflicts. Here, a Polikarpov I-16 represents an aircraft flown by Chinese ace Colonel C.S. Lau, who flew with the Chinese Nationalists against the Japanese in 1938.

Phil Makanna

The debuts of the I-16 in 1998 and the I-153 in 2000 should not be underestimated. These types are a significant addition to the airshow scene. The considerable effort undertaken by Sir Tim Wallis and his staff is vindicated when one is witness to a sight that has not been seen anywhere in the world since the end of World War II.

Ian Brodie

Below: A few days before the show and tractors are set out in their display slots. George Wallis, Don and Shona Robb and their team work hundreds of voluntary hours to present the vintage tractors and machinery that are seen at Wanaka.

Antony Hansen

Ian Brodie

Julian Denmead, Greg Olsen and Graeme Barber of the Warhorse team ham it up. *Ken Chilton*

This is not just an airshow but an occasion! Visitors from Canterbury show their red and black heritage. *Chris Hinch*

Garth Hogan in the cockpit of his Curtiss P-40 Kittyhawk, which debuted at Wanaka in 2000. Garth and his team spent many sleepless nights working to achieve the first flight of the aeroplane, which was just before Easter, to enable the trip south from Auckland. *Chris Hinch*

Another show and another formation — as always, different to the last. For the first time in over 50 years two Kittyhawks are seen together in New Zealand skies, with two Mustangs, a Hurricane and a Spitfire. *Ken Chilton*

Built before man achieved powered flight, this 1889 Shand Mason horse-drawn steam fire engine was purchased for the Ashburton Volunteer Fire Brigade in the same year. Last used in 1937, the machine was stored in an old shed before being completely stripped and rebuilt in 1999. Brigadesman John Newlands then painstakingly built the beautiful working scale model from scratch. *Ken Chilton*

After the AFC's P-40 was crash-landed in a paddock in 1997 it was sold to American Dick Thurman. The aircraft was rebuilt and painted in her original Aleutian Islands colour scheme, just in time to return to Wanaka for the show. Here she displays her dramatic nose beside the P-40 that is owned by Garth Hogan and Charles Darby. *Chris Hinch*

The man who started it all, the man whose enthusiasm for aviation has created what is now considered one of the best warbird airshows in the world: Sir Tim Wallis. *Phil Makanna*

Fiery Finales

Every two years, Alpine Deer Group builder and maintenance man Kevin (Bomber) Harris comes to the fore when he and his team gleefully preside over rockets, bombs and big explosions. Kevin is another one of those behind-the-scenes people who is such an important and integral part of the show.